Revealing the past, informing the future

Revealing the past, informing the future

A guide to archaeology for parishes

Joseph Elders

CHURCH HOUSE PUBLISHING

Church House Publishing
Church House
Great Smith Street
London
SW1P 3NZ

Tel: 020 7898 1571
Fax: 020 7898 1449

ISBN 0 7151 7603 X

Published 2004 for the Council for the Care of Churches by Church House Publishing

Copyright © The Archbishops' Council 2004

Printed in England by Halstan & Co. Ltd, Amersham, Bucks

All rights reserved. No part of this publication may be reproduced or stored or transmitted by any means or in any form, electronic or mechanical, including photocopying, recording, or any information storage and retrieval system, without written permission which should be sought from the Copyright Administrator, Church House Publishing, Church House, Great Smith Street, London SW1P 3NZ; email: copyright@c-of-e.org.uk.

Contents

	List of illustrations	vii
chapter 1	Introduction	1
chapter 2	What is the 'archaeology' of parish churches?	3
chapter 3	When to seek advice and what to look out for	6
chapter 4	The legal requirements	30
chapter 5	The mitigation of archaeological implications	36
	Further reading	41
	Organizations able to give advice	43
	Index	44

It is important to remember that a faculty will be required before any work can be undertaken in a Church of England church. Contact your Diocesan Advisory Committee for advice.

List of illustrations

Fig. 1 The church of Bradwell-on-Sea, St Peter, in the 1930s   7

Fig. 2 The top two stages of the tower at Little Livermere, St Peter, were added in the eighteenth century to make the church a landscape feature   8

Fig. 3 Dendrochronological investigation at the timber-framed church of Denton, St Lawrence   13

Fig. 4 Malmesbury Abbey stone screen   15

Fig. 5 Medieval wall paintings in the nave of Troston, St Mary, depicting St Nicholas and St George   17

Fig. 6 A Saxon carving built into the wall and modern drainpipe at Sherburn, St Hilda   19

Fig. 7 Gravestone at Great Livermere, St Peter, to the (Falconer) of King Charles I and II and James II. This person's skeleton could tell us a great deal when linked to this knowledge of his life   20

Fig. 8 Fourteenth-century effigy of a lady, decaying in the churchyard at Weaverthorpe, St Andrew   21

Fig. 9 Scratch dial in a wall at Wintringham, St Peter   22

Fig. 10 The churchyard at Fulbourn, St Vigor with All Saints, where a gap in the gravestones marks the location of the demolished second church in the churchyard   26

Fig. 11 The ruined church at Arborfield   34

chapter 1

Introduction

England's historic parish churches have been used for Christian worship for up to 1,400 years. This continuity of use has protected and preserved them, so that the parish church is almost always the oldest building in a village or town, and together with its churchyard often forms an island of ancient remains within a constantly changing environment. Churches are, of course, much more than islands of history – they serve as a centre for local worship and mission, an icon of community memory and a focus for social activity.

However, churches and churchyards are rich in resources for understanding the past and have huge research potential, not only for the archaeologist but also for everybody interested in local and national history. Church buildings are by no means frozen in time but have evolved over the life of the parish, which makes them all the more important and fascinating. Changes to the fabric of the church building are not always easy to identify, and archaeological work can help us to understand how and why a church has developed as it has over time.

Of course, some churches are outstanding examples of their time and possess a special ambience or exceptional features and furnishings which careless change can destroy. Nobody is interested in preventing change at any price, but change must be responsibly managed if this fragile inheritance, the material evidence of our national Christian witness, is not to be lost. If the archaeology of a church or its churchyard has to be disturbed to allow necessary works, it must be recorded in full to add to our information about the past, and to aid our understanding for the management of the church building and churchyard in the future.

Although archaeology is a specialist discipline, it is one in which members of the local community can take a great interest. Significant archaeological work at the parish church may attract the curiosity of local people, and in some cases it may be possible for enthusiastic amateurs to get involved – for example, by helping to record the inscriptions on gravestones. Elderly members of the parish may remember changes being made to the building even if there are no records of them.

The more completely we understand the history and archaeology of a church, its churchyard and the surrounding environment, the greater our appreciation becomes of why a particular place is unique. On a more everyday level, understanding the history of a church also helps a parish to recognize when proposed works of maintenance or development may have archaeological implications, and thus reduce delay, cost and damage to this inheritance. Understanding the building will also help us to achieve responsible and cost-effective management, avoiding damaging and potentially expensive mistakes.

chapter 2
What is the 'archaeology' of parish churches?

The double meaning of the word 'archaeology' as used by archaeologists must first be explained, as this can cause confusion. The word is most properly used to describe the discipline of archaeology, as one might define an academic department, such as a department of Geography, English Literature or Biology. It is also used as shorthand for the material remains themselves which archaeologists study, hence one speaks of the 'archaeology' of parish churches. The Council for British Archaeology (CBA) defines this as:

> the complete historical study of the fabric and material remains of a church, above and below ground, in relation to its site, contents and historic setting and to its community.

This sentence is worth reading again, as we are dealing here with significant definitions. Archaeology is often perceived as a heads-down discipline, to do with excavations of buried remains such as pits, ditches, wall foundations, burials and pottery, carried out with spade, trowel, brush and bucket. This perception has been reinforced by the recent spate of television programmes on the subject, almost all of which deal with below-ground archaeology, particularly the excavation of human remains (of which more later).

The discipline of archaeology has always been much more holistic than this, embracing the landscape, the built environment, historic objects and living (and worshipping) communities, employing the same analytical techniques based on observing changes in material culture to interpret the past. A relocated oak screen, a fragment of a ledger slab, the slots for the lost rood loft or a blocked lancet window in a chancel wall are archaeological 'finds' just as much as a burial vault or the underground remains of earlier structures beneath the present building.

Seen in this way, the archaeology of a church and churchyard encompasses everything within its curtilage, from the boundary (or boundaries, as these have often moved) and gates to the churchyard, buildings and monuments within, and from the top of the church spire to the bedrock beneath. It will encompass the entire history of the site, including its pre-Christian use in some cases. Even nineteenth- or twentieth-century churches will be of interest to the archaeologist, who may be asked to record architectural or art-historical features or objects prior to structural changes, excavate burials, or investigate the remains of earlier settlement on the site, even on supposedly 'new' sites.

The archaeology of a church will also encompass the movable and immovable fittings and furnishings and the traces of their former existence, including evidence for former chantry chapels, rood screens, wall paintings, altars and much more. All of these are of interest in their own right but also tell us much about the history of liturgical practice and popular devotion. Christianity is a dynamic faith, and the changing practices associated with it throughout its long history can often still be traced in our churches today using archaeological techniques.

The church archaeologist will also look outwards beyond the churchyard boundary, and appraise the building and site in its context. Our churches and churchyards are part of the broader historic environment, a part of all our lives which we tend to take for granted but which to a large extent provides us with our sense of identity and enriches our collective and individual consciousness and spirituality. The need to understand and appreciate our historic environment and our collective past is therefore not merely an academic pursuit, the domain of eggheads in ivory towers, but of profound importance, influencing our notions of nationality, community and self. This is the context in which the work of church archaeologists must be seen, and in which the resource they seek to protect and study must be valued.

Church archaeology is changing fast. In recent years, there has been a considerable increase of interest in it, as exemplified by the foundation of the Society for Church Archaeology in 1996. There have been significant advances in the scientific techniques available to archaeologists, such as dendrochronology (tree-ring dating), DNA analysis, fabric surveying and recording techniques, and aerial photography or geophysical survey to reveal traces of graves, buried walls or vaults no longer visible at ground level. This has all led to an increasing number of interested parties who, in one way or another, could benefit from archaeological work in churches. Some may wish to add, however: 'yes, and those who suffer when the bills are settled'.

The fact that this comment is often heard means that in most cases there has been a failure somewhere along the line, most likely of communication. The rest of this booklet is devoted to attempting to explain how the benefits can and should outweigh the costs, and how best to maximize the former and minimize the latter.

chapter 3
When to seek advice and what to look for

A common question, and an important one, is: When should one seek advice about the possible archaeological implications of any work to be carried out? The simple answer is: as early as possible. An exhaustive list of works to a church or churchyard which may have archaeological implications is impossible, but common types include:

- building an extension to the church;
- creating facilities such as toilets or kitchens within the church;
- erecting a free-standing building within the churchyard;
- repair work, for example replacing stonework, timbers or plaster;
- digging drains or laying cables or pipes;
- installing internal lighting or floodlighting;
- installing telecommunications equipment;
- installing or repairing heating systems;
- inserting features such as aumbries, galleries or suspended floors;
- creating a new area for cremated remains;
- landscaping;
- digging of new graves in an area clear of memorials, but perhaps not of burials;
- repair or conservation work to monuments;
- laying new paths or repairing existing ones;
- repairing the churchyard wall or building a new wall;
- repointing or replacement of masonry;
- reordering involving floor replacement or the movement of historic fittings;
- repairing a lychgate;
- planting trees;
- clearing grave markers.

The next question that is often asked is: What do archaeologists look for, and what might they find? Again, a comprehensive list of the kind of things that archaeologists have an interest in is impossible, but an attempt will be made here to give some idea,

fig. 1
The church of Bradwell-on-Sea, St Peter, in the 1930s

beginning with the fabric of the church itself and then moving on to its contents and surroundings.

The fabric of the church building

The single-cell church is the simplest form of early Christian church. The two-cell church of nave and chancel is the norm for Anglo-Saxon and Norman churches, but many such early foundations may not always have been the simple boxes they now appear to be. An example of this is the originally seventh-century chapel at Bradwell-on-Sea, Essex (see fig. 1). The simple appearance of this church belies a complex development, the Anglo-Saxon church having had an apse at the east end and symmetrical porticos which are only evident by a careful study of the above-ground fabric and by below-ground excavation and geophysical survey.

Medieval churches expanded and also contracted from the two-cell core, acquiring often a tower, a north aisle, then perhaps a south aisle, but often enough losing one or both of these over the centuries. Chantry chapels were added and often demolished or

fig. 2 The top two stages of the tower at Little Livermere, St Peter, were added in the eighteenth century to make the church a landscape feature

converted after the Reformation, transepts and porches were added and sometimes removed. Windows and doors were inserted and blocked as architectural fashions and liturgical practices changed. In the Victorian period, many churches were partially or wholly rebuilt, and porches, vestries and organ chambers were added. Fragile traces of these changes often remain in the standing fabric which can be revealed by careful recording. These traces can be destroyed by stone replacement or repointing.

An example of the complex development of a medieval parish church is Holy Trinity, Lambley, in the Diocese of Southwell, where the sum of several episodes of relatively minor archaeological recording, observation and excavation has added considerably to our knowledge and understanding of this fine building, and informed its future management. The earliest fabric of the church is the twelfth-century lower stage of the west tower. The core fabric of the nave, south porch and chancel is fifteenth century, but the north chancel wall survives from the fourteenth. A Victorian vestry now blocks much of the evidence in the fabric for the existence of a former two-storey chantry chapel of c.1340 on the north side of the chancel, but a hagioscope from the upper room survives, giving a view into the chancel. A fifteenth-century north doorway to the nave has been blocked up.

During repairs in 1984, several inlaid medieval floor tiles were discovered underneath boards supporting the lead of the parapet gutters on the south side of the nave roof. An evaluation excavation (see p. 38) was undertaken in advance of a proposed extension on the north side of the nave adjacent to the blocked north doorway. Evaluation trenches revealed that archaeological deposits were present at all depths. Late twelfth-century domestic pottery was found along with environmental evidence of domestic settlement. Close to the present north wall of the church the structural remains of a probable former north aisle, post-dating this settlement phase but pre-dating the fifteenth century, were located. This knowledge helped the parish to reconsider its plans.

As at Lambley, the tower is often one of the most archaeologically interesting parts of the church building, frequently being the only medieval structural feature that Victorian restoration left relatively intact. Having said this, stages were often rebuilt or added, spires erected, repaired or dismantled, windows and doors blocked or inserted. Many of the surviving gargoyles and waterspouts are to be found at this high level.

The distinctive round towers of East Anglian churches are justly famous, if not all Saxon in date and defensive in nature as was once thought. Many of the fortified church towers of northern England were most definitely defensive, with their fireproof stone roofs, raised doorways, arrow slits and massive walls. Many towers were heightened in the eighteenth and early nineteenth centuries to enhance their value in the landscape designs of that period (see fig. 2). All these changes of use and design are of interest, and often traceable by the archaeologist even after later alterations have obscured them.

Bells, bellframes, belfries and clocks

England possesses in its church towers a quite unique resource, its historic clocks, belfries, bells and bellframes. Any works to these, even if apparently routine repair works, should be carefully considered, and advice sought from the Diocesan Advisory Committee (DAC). Both the DAC and the Council for the Care of Churches (CCC) have specialist advisers, and the CCC holds a national list of historic bells which can be consulted.

Most dated and inscribed bells will date from the post-medieval period, but an increasing number of pre-Reformation examples has been recognized. The oldest tower bell in Great Britain is at St Botolph, Hardham, still in use but dated before 1100. The oldest inscribed bell is the Gargate bell at St Lawrence, Caversfield, which is dated c.1200–10. The oldest dated bells are early thirteenth century, for example the bell at St Chad, Claughton, which is the only medieval object to survive at this church.

Historic turret clocks are likewise of great interest and often of high quality. In the seventeenth century Thomas Tompion, often called the father of English clockmaking, made turret clocks for churches, as did his famous contemporaries Joseph Knibb and William Clement. One of Clement's clocks can be seen in King's College, Cambridge, and there are two turret clocks by Knibb at St Mary the Virgin, Oxford. Any clocks of this date or earlier will be of the utmost rarity and importance. Clocks and their mechanisms may be integral to the fabric of the tower, and recording should be undertaken before removal or repair. Similarly, the installation of new parts, dials or new clock mechanisms may require penetrating the fabric, and this should be avoided or, if this is not possible, recorded.

The study of bellframes and other wooden objects and structures has been revolutionized in recent years by the technique of dendrochronology, better known as tree-ring dating. This has not only proved that some structures are much older than previously thought, but also that the reverse is sometimes true. The oldest bellframes seem to date from the fourteenth century. Dendrochronological dating of the bellframe of the priory church of St Bartholomew the Less, London, in 2003 suggested felling and construction in the years 1481–1507. This is some 70 years earlier than was surmised on stylistic grounds alone. Stylistic evidence may be present in the form of pre-Reformation features such as curved side frame braces, king posts or short heads, but many bellframes have been altered so many times that they may incorporate older material in rebuilt frames, or little historic material may survive after constant replacement. With later examples the date might be known or obvious, such as is the case with the inscribed frame of 1626 at St Michael, Onibury, but this is rare. Advice should always be sought from the DAC.

It is indeed often the case that bellframes have survived all restoration work by later generations, but many are near the end of their useful lives. Even if this is so, historic material should be recorded. The bellframe at St Mary, Lowgate, in Hull was built in 1727 and was rebuilt at the top of the tower in 1860, since when the vibrations from bell-ringing have endangered the stability of the tower, and it had to be removed. It was carefully recorded by archaeologists before this was done. Not only bellframes, but sometimes whole timber belfries survive, as at All Saints, Doddinghurst. This fourteenth-century belfry has been dated and recorded archaeologically, and shown to be an independent, fully framed structure that sits within the flint walls of the church. Timber supports to ringing chambers may survive, often post-medieval in date as at Romsey Abbey or at St Michael, Aylsham, where a rare medieval example survives.

Roofs, spires and structural timbers

It is quite extraordinary how many historic roofs have survived in parish churches, a testimony to the strength and endurance of English oak and the skills of the carpenters and craftsmen who built them. Great advances have been made in the technical study and dating of roofs in recent years, the latter mainly due to the development of the technique of dendrochronology mentioned above, but also to the study of carpentry techniques pioneered by Cecil Drewett in the 1960s. These techniques have enabled us to establish that many church roofs are medieval in origin, with some dating back as far as the twelfth century.

The oak roof of St Mary, Kempley, is perhaps the oldest timber roof in Britain, dated by dendrochronology to 1120–50 (also the date of the famous wall paintings at this church), with medieval and post-medieval repairs. Clearly, as in this example, if advice is sought before repairs are effected, and the necessary works are recorded under archaeological supervision, it will be possible to manage the structure much more effectively, as well as increasing our knowledge of the building.

Occasionally even the wooden battens covering the rafters may survive for centuries, as shown by the recent discovery of thirteenth-century examples at Salisbury Cathedral. This shows that extreme care must be taken when repairing historic roofs, as much more ancient material may survive than is expected. This is also true of the covering material, whether lead, clay tiles or shingles, and of the rainwater goods.

Wood was also used for spires, for example for the famous crooked spire of St Mary, Chesterfield, or the Sussex shingled spires, such as at Holy Trinity, Bosham. A study of this spire in 1998 showed that although the mast and cross trees that support it were replaced in 1841, the remainder of the complex structure is essentially of one build, constructed of timbers felled in 1405–6.

Timber was much used in many parts of the country where there was a shortage of good building stone, such as Essex, where there are a number of notable timber porches and the only timber stave or log-built church in Britain, at St Andrew, Greensted-iuxta-Ongar. This used to be considered Anglo-Saxon after a tree-ring date of 845 was calculated, but further work now indicates that it was probably built just after the Norman Conquest in 1066. There are also around 30 known timber-framed churches, such as the group in the north-west to which St Werburgh, Warburton, belongs. While it may be considered unlikely that many more such examples will be discovered, remnants of timber framing may still survive within church walls and porches.

Floors

Many churches had simple earthen floors strewn with rushes until as late as the eighteenth century. Where they survive, early floors can be helpful in showing how the interior of the church used to be laid out, including where, for example, fonts, altars, rood screens or shrines once stood. Early floors are also important to

fig. 3
Dendro-chronological investigation at the timber-framed church of Denton, St Lawrence

the archaeologist in helping to relate different periods of building activity in different parts of the church, and occasionally by producing artefacts which help to date them. Sometimes medieval stone-flagged or tiled floors will survive, sometimes in patches in an existing floor, perhaps embedded within Victorian copies, or buried beneath a later floor. Later floor materials including pammets, flagstones and also Victorian tiles can be of high interest and quality, and they may contain ledger slabs and memorial brasses (see p. 21). Disturbing such floors, for example during pew removal or repair or the insertion of underfloor heating systems, can damage this fragile evidence, and may disturb graves and vaults. Existing heating systems, including iron grilles or fixtures, may be Victorian or part of the historic interest of the building in their own right, and should be retained if possible or recorded before removal.

Furnishings and fittings inside and out

Returning to St Mary, Kempley, this church possesses other historic woodwork, which has also been investigated using dendrochronology. As well as the roof, both the south door of the nave and the west door into the tower appear to be primarily Norman carpentry. There is also an oak chest in this church which has yielded a felling date range of between 1492 and 1522. Many medieval chests are lying in the vestries or vestibules of churches throughout the country, often unappreciated. The new techniques available may well help to date many such wooden items.

Other types of wooden furniture can also be of high archaeological potential. Examples are benches and pews, the earliest examples of which date back to the late thirteenth and fourteenth centuries, as at St Mary the Virgin, Warmington, St Mary the Virgin, Gamlingay, and All Saints, Dunsfold. Georgian box pews and Victorian and later pews are often of high quality and interest in themselves, and pews, wall panelling and screens may also contain medieval or post-medieval elements.

There are a surprising number of medieval pulpits and, more commonly, of restored pulpits which incorporate medieval material. Much more common, but nevertheless of great interest, are post-Reformation pulpits, reflecting the shift in emphasis from altar to pulpit at this time. Many altar tables and chairs also survive from this period, often standing unused in aisles or vestries. Some churches still have their full complement of historic furnishings and fittings from a particular period, such as the Georgian schemes at All Saints, Skelton-in-Cleveland, and St Mary, Whitby, with their galleries, box pews and three-decker pulpits, but archaeological evidence will often survive at churches which have seen more changes since this period.

Timber screens form another major category of historic woodwork that is not only of interest in its own right, but casts a great deal of light on the development of liturgy and popular devotion. The late thirteenth-century screens at St Mary's Hospital in Chichester are among the best preserved. The majority of surviving screens date from the fifteenth century, such as the example from the beautiful

When to seek advice and what to look out for

fig. 4
Malmesbury Abbey stone screen

church of St Peter, Wintringham. This is particularly interesting, with its tiny cut-out squint with cusped head which allows a view of the altar from a kneeling position. At least as important are the rare examples of stone screens, such as the famous example at Malmesbury Abbey (fig. 4). When such screens are relocated or dismantled, their context and liturgical significance can be lost.

Rood screens are a special category, with their related stairs within the thickness of the chancel walls and the redundant slots in the chancel arch. The earliest surviving rood screen is the superb example at St Michael, Stanton Harcourt, dating from the thirteenth century. The door has its original bolt, lock and fittings, a rare detail. Many screens and all roods were destroyed during the Reformation and the later Puritan campaigns, but some parts survive, mostly in fragments reused elsewhere, whose origins are not always recognized.

Galleries were often added to churches in the eighteenth and early nineteenth centuries, and these are of interest whether they are intact from this date or if only residual timbers, slots or other

15

features survive. Occasionally wood from the rood screen or other timbers was reused to create these, and there are occasional examples of earlier galleries.

There may also be evidence for the existence of musical instruments, particularly organs. Contrary to popular belief, there is a very long tradition of the use of organs in churches, going back to at least the tenth century in England. A famous organ was built c.950 for Winchester Cathedral, but the earliest survivals are much later, for example the early sixteenth-century case at St Stephen, Old Radnor, and the late seventeenth-century chamber organ at St George, Nottingham. The pipework, machinery and the cases of organs are of historic and archaeological significance and potential, and this should be considered when repairs or changes are contemplated.

The position of integral features such as squints and piscinae can also be of interest, not only in themselves as architectural features but also for what they tell us about liturgical arrangements in the past. Their presence will signal the existence of side chapels and altars. Some will have been plastered over or moved. The same is true of aumbries, while fonts have very often been moved from their original position, sometimes standing ignored and unused in tower spaces or storage rooms, occasionally even dismantled. Parts of or even whole fonts, particularly post-Reformation ones, are occasionally found buried in churchyards, and sometimes even in gardens, serving as birdbaths or flower stands. Mensa slabs and other altar fragments can similarly be found discarded, and occasionally parts of pre-Reformation statuary and monuments, too.

Reredos can also be of potentially great archaeological interest, even though most will be the product of the Victorian revival, and recording of these, and later examples, may be necessary if they are altered or removed. Few earlier examples survive, but more are still being discovered; they will be of the highest interest. At St Mary, Brent Eleigh, a painted reredos was revealed during construction work in the 1960s, a rare example actually painted on the wall behind the high altar. It probably dates from the early fourteenth century and is of a high quality comparable to the contemporary retable at St Mary, Thornam Parva. Such survivals often have a long and chequered history behind them. The retable at Thornam Parva probably came from the Dominican priory at Thetford and survived the Reformation, perhaps being used for Catholic worship

When to seek advice and what to look out for

fig. 5
Medieval wall paintings in the nave of Troston, St Mary, depicting St Nicholas and St George

by a recusant family living on the estate in the seventeenth or eighteenth centuries. It was found in the nineteenth century in a barn and was donated to the church in 1927, where it can be seen. Similarly, the late fourteenth-century Despenser reredos in Norwich Cathedral was saved from destruction over the centuries by its use until 1847 as the underside of a table. There may be many more such artefacts, built into walls, stored away or fulfilling other functions.

As the example of Brent Eleigh discussed above illustrates, medieval wall paintings are being discovered all the time, belying the misconception that the Reformation destroyed all such art in churches in England. There are remains of hundreds of medieval wall paintings in English churches, and more are discovered each year – often hidden over the centuries behind layers of plaster (see fig. 5). Restoration is currently underway on the recently discovered wall paintings at St Mary, Houghton-on-the-Hill, where the paintings are perhaps the earliest found in modern times in England. The survival of the paintings is remarkable, due, as in many cases, to a combination of good fortune and care, which should always be taken when plaster is stripped or wall monuments moved. Again it should be emphasized that it is not only medieval wall paintings that are of interest. Post-Reformation texts and indeed Victorian paintings, schemes and texts are also important, and are likewise often hidden behind plaster, monuments or furniture, or simply overlooked.

It is fairly rare now to discover unknown examples of ancient stained glass in church windows, but it does happen, and as with fonts there are occasional finds of broken fragments when excavations are undertaken within the curtilage of a church. This happened at St Peter, Wintringham, where fragments of glass were found under the floor of the church where they had been buried after the Reformation. They have since been restored and remounted in the medieval windows of the church.

Monuments

The funerary monuments inside and outside a church are an integral and important part of its archaeology. They may indicate far-flung contacts and patronage, hint at pilgrimage and trade routes, commemorate wars and colonies abroad, and speak of crafts and industries long since gone. There may still be family connections with the local community; indeed, it is often surprising how many of such links have survived into the present time, especially in rural areas. When such monuments are destroyed or moved, a wealth of information and layers of meaning can be irretrievably lost. Such losses can be avoided through care and forethought. Even in the case of relatively minor and inexpensive works for which a faculty might not strictly speaking be necessary, it is always a good idea to ask for the freely given advice of the DAC.

A simple example of this occurred recently, where an Archdeacon's licence was granted to clamp an early thirteenth-century grave slab to the sanctuary wall. No faculty was considered necessary, as the cost involved was very small. However, the DAC was asked to advise, and was able to inform the parochial church council (PCC) that such clamping can be deleterious not only to the monument, leading to spalling and cracking, but also to the church fabric, trapping dampness onto the wall behind. In this particular case the slab turned out to have been carved with crosses on both sides, one of which would have been hidden and perhaps damaged, if not lost forever, if the work had gone ahead. Such cross slab fragments were

fig. 6
A Saxon carving built into the wall and modern drainpipe at Sherburn, St Hilda

often built into walls, mostly by Victorian restorers. A particularly dramatic example is the case of St Brandon, Brancepeth, where whole medieval grave slabs were discovered within the fabric of the walls during restoration work following the catastrophic fire there.

More rarely, fragments of Anglo-Saxon monuments and sculpture might be found in the church walls (see fig. 6) or buried in the churchyard, for example parts of large, free-standing stone crosses, or parts of so-called 'hogback' grave covers. Stone replacement is a particular danger to this kind of rare and fragile artefact, as they might not be recognized, and may be built with the decorated side within the wall.

Gravestones, monuments and tombs can give us both archaeological and documentary evidence (from inscriptions) about the history of a church. Some of these monuments may be so significant that they are listed in the same way as important buildings, and fine examples can be found in many ancient churchyards. It is a good idea to record such monuments as part of a wider churchyard plan (see p. 23). It should be noted that conservation and repair work to monuments, especially to larger examples and table-tombs, can have archaeological implications

fig. 7
Gravestone at Great Livermere, St Peter, to the (Falconer) of King Charles I and II and James II. This person's skeleton could tell us a great deal when linked to this knowledge of his life

in the same way as work to the fabric of the church itself. The Diocesan Archaeology Adviser (DAA) should be informed of any planned conservation work to such tombs.

The relationship of monument or grave marker to the interred individual is of prime importance to the archaeologist and historian. If we know who a particular person was, how old they were when they died, their lifestyle and their family relationships, this is of interest not only to the student of genealogy but also to the social historian and the pathologist (see figs 7 and 8, and the section on the interment of human remains on pp. 27–9). Important information can therefore be lost when headstones within the churchyard are moved without record, and often piled or lined up against the church wall. Not only is the original context lost, including the knowledge of the very existence of burials of a certain date in that place if no record is made, but again this traps damp and encourages vegetation growth against the wall.

fig. 8
Fourteenth-century effigy of a lady, decaying in the churchyard at Weaverthorpe, St Andrew

Monuments inside churches

Movement of monuments within a church should also be considered carefully. Some wall-mounted tablets have been moved from their original positions, as have many monumental brasses, ledger slabs and effigies, though this should not be assumed to be the case without research. Study of the important effigies at the church of St Margaret, Ifield, proved that the assumption, often made, that their present locations date back to Victorian reorderings was wrong. There are indeed still a great number of monuments which do mark the original place of interment. This should be borne in mind when undertaking reordering of interiors or relaying of floors. If monuments must be moved, their original position should be recorded.

fig. 9
Scratch dial in a wall at Wintringham, St Peter

The cleaning and conservation of monuments can also have archaeological implications. Temporary removal for cleaning or repair can reveal details of design and construction which should be recorded, and such action may also reveal features of interest behind or under the monuments themselves, such as vaults, blocked windows or wall paintings.

Sundials

A category of artefact often overlooked is old sundials, either built into the church walls or often remounted above doors and porches (see fig. 9). These are very susceptible to damage by stone cleaning and replacement. The DAA will be able to give advice on all such works through the normal processes of DAC consultation, even where only like-for-like replacements are undertaken.

The churchyard plan

Where a general reordering of the churchyard is proposed, a plan of all monuments and stones, correctly to scale, should be prepared; this should be provided with a full copy of all the inscriptions, numbered to correspond with the plan. The inscriptions on older and more weathered memorials can be difficult to read, and it is often essential to enlist the help of experts. Apparently lost portions can sometimes be read by the use of oblique light (e.g. from a torch) or in some cases laser analysis, or can be reconstructed by careful comparison with the registers.

It is of particular importance that all inscriptions dating from before about 1850 should be recorded carefully, as these in almost all cases provide a source of information not otherwise collected (e.g. actual date of death) before civil registration began in 1837. Members of the Society of Genealogists, the Federation of Family History Societies and members of local history societies are often able to assist parishes with the compilation of descriptive lists of churchyard memorials.

There will be many old graves and vaults whose positions cannot be identified simply by looking. Archaeological evaluation in the form of geophysical analysis and, where appropriate, test excavation may help to identify these in advance of any works, although mounds and depressions may also yield clues as to their location. These should all be plotted on the churchyard plan along with grave markers, monuments, kerbstones and areas set aside for cremated remains, plus, of course, the church itself. The boundary features of the yard such as hedges, fences, trees and gates need to be included, as do paths, both hard-surfaced and mown grass. It may not seem important today to have such accurate detail, but it may pay dividends to the next generation who take over the management of the churchyard. This record should be deposited in the Diocesan Record Office. Archaeology students from a local university or members of a local history or archaeological society may be willing to help compile a churchyard plan.

The churchyard and its historic environment

Research is still going on to establish just how old our earliest churchyards actually are. It is clear that, in most medieval parishes, churchyards were in existence by the eleventh century, and the majority of these are likely to be older than this. The evidence for these dates comes from a variety of sources, including the study of the topography of the area and historical documents, and from direct evidence such as human remains, grave markers and finds from archaeological excavations which can now be investigated using sophisticated dating techniques.

Churchyards were not always the tranquil places they generally are today. In the medieval period, some were used for commercial activities. Markets and fairs held within a village or town often occupied the open area around the church and left traces which sometimes survive in the form of pottery and other goods. For example, until 1223 there were apparently skin and cloth markets in the churchyards of St Peter and St Michael, Lincoln. Activities such as bell-casting were frequently undertaken in the churchyard prior to the days of effective transport, and traces of construction activity such as pits for the mixing of mortar or for slaking lime frequently occur near a church.

There may be traces of earlier buildings within the churchyard. The present church may not be the first on the site and may not be directly related to earlier buildings which could lie elsewhere within the enclosure. There are a number of churches built on Roman sites, like St Kyneburgha, Castor, or St Mary and All Saints, Rivenhall. There may also have been free-standing towers and ancillary buildings. Sometimes, as at Asheldham in Essex, the priest had a house within the churchyard. Foundations, floors or graves may survive from an earlier building or an earlier stage of the existing church. Immediately adjacent to the present external walls will almost certainly be fragile evidence below the turf which relates to the construction of the building and to any lost portions such as aisles, chapels or porches, evidence which is particularly at risk from well-intentioned attempts to improve the drainage.

Supposedly 'new' sites can also have archaeological potential. The site of the 1960s church of St Edmund the Martyr on Temple Hill, Dartford, a church built to serve a new housing estate on what was previously farmland, is of great archaeological significance due to the presence of a large Iron Age and Romano-British cemetery on the site. Though the name and the attractive hilltop location of the site overlooking the River Thames were clues to the existence of possible archaeological remains in this case, there will not always be such pointers. Contact your local authority's Sites and Monuments Record (SMR). It is worth noting that at the time of writing, many of these were being upgraded and renamed Historic Environment Records (HERs).

Traces of earlier buildings or occupation, construction activity and commercial trading are not the only type of archaeological evidence to be found. The primary function of churchyards was burial, and this will be reflected in the presence of a large number of graves, many more than are represented by surviving headstones. Graves have often been disturbed or cut by later burials, especially the medieval ones which were sometimes quite shallow. It is important to realize that human remains can occur within a few centimetres of the modern land surface, both disarticulated individuals and complete burials (see the section on the treatment of human remains, p. 27).

Looking beyond the churchyard, the church may stand away from the modern village or at an important town crossroads; it may dominate the area from the highest point, be tucked away down a rough track, or shoe-horned into a tight urban plot. The churchyard may have been encroached upon, or been extended at various times. There may be the remains of several houses which at one time encroached on the churchyard – or indeed the reverse, as happened at All Saints, Crondall. Archaeological recording in the churchyard in 1997 revealed that the cemetery had been extended in the eighteenth and nineteenth centuries over the remains of seventeenth-century buildings. Depressions, tree lines and ridges may hint at earlier boundaries. There may once have been two churches in the same churchyard, serving separate manors, as at Fulbourn in Cambridgeshire, where the only clue today to the second, vanished church is a rectangular area with no grave

fig. 10
The churchyard at Fulbourn, St Vigor with All Saints, where a gap in the gravestones marks the location of the demolished second church in the churchyard

markers on it (fig. 10). All of these things will contribute to our understanding of the development not only of the church but also of its surrounding countryside, buildings and community over the centuries.

Objects and finds

The main concern of archaeologists when excavating a site is the evidence of structures and their relationships which helps to reconstruct the past. In association with these physical artefacts may be found such things as coins, pottery and tools, which help both to date and to interpret the archaeological layers. Any such finds are the property of the incumbent, but during a vacancy the PCC has a proprietary interest in them. Although rarely of any monetary value, they may have historical and local significance. Casual finds, made for example whilst digging graves, may also be of relevance to the history of the church and churchyard and should be reported to the County Archaeologist, the DAA, the SMR, and to any established local museums. It is important to conserve any significant finds to prevent their decay and to map the discovery on the SMR. Individually, finds may appear trivial, but when plotted regionally they often reveal patterns.

All finds are subject to the faculty jurisdiction and therefore a faculty must be obtained before any substantive work (other than recording) is carried out on them or before they are removed from the site. The Church is negotiating an exemption from the Treasure Act (1996) whereby finds from graves within churchyards will not come under the jurisdiction of the state but this is not yet in force. This brings us to one class of 'object' or 'find', very common in and around churches, that is a special case: that of human remains.

The treatment of human remains

> Good friend, for Jesus sake forbear
> To dig the dust enclosed here
> Blest be the man that spares these stones
> And curst be he that moves my bones.
>
> *Epitaph on William Shakespeare's grave marker*

The quotation above encapsulates an attitude which would appear to be peculiar to post-Reformation England: the wish for eternal peace to mean just that for the mortal remains of the deceased. It is a wish which should be respected, but which cannot always be strictly adhered to. Nevertheless, disturbing burials should never be undertaken lightly. When the decision has been taken, archaeological recording will often be required.

The CCC report *Church Archaeology: Its Care and Management* (1999) identified the care and treatment of human remains disturbed during archaeological work as one of the most complex, difficult and emotive issues in church archaeology. Human remains comprise one of the commonest forms of 'find' recovered by church archaeologists in a variety of circumstances. The report made the following recommendations in relation to this issue which summarize the legal position under the Faculty Jurisdiction:

- There should be a presumption against the disturbance of human remains.
- Disturbed remains should be afforded respectful treatment.
- There should be a presumption in favour of reinterment of remains.

It is very likely that an excavation (whether archaeological or otherwise) within a churchyard will reveal human remains. While care should always be taken to avoid damage to all archaeological deposits, in the case of human remains the reasons for preservation *in situ* go deeper than simply a desire to preserve information about the past. Pastoral, ethical and theological sensitivities need also to be taken into account. The successful public outcry over the insensitive exhumation methods proposed for the clearance of the cemetery at St Pancras Old Church, London, for the construction of the Channel Tunnel Rail Link is a good example of this.

The explosion of public interest in archaeology and the concurrent flood of television programmes such as Channel 4's *Time Team* and, in particular, BBC's *Meet the Ancestors* has highlighted our fascination with the past and, specifically, with its most human dimension, our ancestors, the dead. Archaeologists find themselves under pressure from conflicts between scientific interest, public fascination and an often unspecific and poorly articulated wish to show respect to the dead.

The anxiety often expressed when human remains are exposed can be allayed through the sympathetic treatment and careful removal of the remains. If human remains are to be disturbed during an excavation, good archaeological practice dictates that this should be carried out with due respect, preferably behind screens. Sometimes, but by no means always, important information can be obtained through the scientific study of human remains revealed during an archaeological excavation. The bones may, for example, preserve evidence of disease, commonly arthritis, syphilis, rickets or osteoporosis, with all the possibilities this line of inquiry opens into the effects of certain lifestyles, professions and diets. The new possibilities afforded by the study of DNA from ancient bones are promising but need to be handled carefully, to avoid abuse for political or trivial motives.

If a case can be made for removing remains from the site for further study, they should eventually be reinterred along with any coffin fittings or other grave goods, unless a very convincing case can be made for an alternative course of action. This will ultimately be at the discretion of the Diocesan Chancellor. The Cathedral and Church Buildings Division and English Heritage have produced guidelines on this difficult issue, found on the Church of England web site or available from the DAC.

chapter 4
The legal requirements

The faculty jurisdiction of the Church of England dates from the time of King William I, 'the Conqueror', who prescribed that all cases relating to the cure of souls be tried in the church courts. The necessity of obtaining the sanction of the Ordinary for changes to church buildings was asserted by a constitution of Otho, the Legate in England of Pope Gregory IX, made in a national synod in 1237:

> We strictly forbid ... Rectors of churches to pull down ancient consecrated churches without the consent and licence of the Bishop of the Diocese, under pretence of raising a more ample and fair fabric. Let the Diocesan consider whether it will be more expedient to grant or deny a licence.

This can plausibly be credited as the first buildings control legislation in England, several centuries in advance of secular legislation. The faculty jurisdiction has developed along with the latter, since the secular authorities began to catch up in 1882 with the introduction of the Ancient Monuments Act. The dioceses now have their own Archaeological Advisers (DAAs) who give advice through the DAC. The DAAs come from a variety of professional backgrounds, but many are local authority archaeologists. The DACs are coordinated by the CCC at national level, now as part of the Cathedral and Church Buildings Division of the Archbishops' Council. A PCC now has the right to ask the CCC directly for advice, as has the DAC and the Diocesan Chancellor.

The so-called 'Ecclesiastical Exemption', under which the faculty jurisdiction works, exempts Anglican church buildings from Listed Building control and Scheduled Ancient Monument consent, but this does not mean that secular authorities are never involved with the management, conservation, protection and development of our historic churches and churchyards. If significant changes to listed churches and their churchyards are planned, a large number of bodies may need to be consulted. These are English Heritage (currently only Grade II* and Grade I churches: see *New Work in Historic Places of Worship*, English Heritage, 2003 for details) and the relevant Amenity Societies (see the list of organizations able to give advice, on p. 43). While this is often seen as a burden in terms of time, expense and bureaucracy, the involvement of experts from a wide spectrum of interested parties can often help parishes to find cheaper and better solutions, and to make better use of their buildings. It also avoids unwelcome (and possibly expensive) surprises later on when one of these bodies registers opposition to the proposals.

Any significant changes to a church or churchyard will require a faculty from the Diocesan Chancellor before they can go ahead, acting on advice from the DAC and the other bodies mentioned above where appropriate. The Ecclesiastical Exemption does not exempt the Church from planning permission, which means that significant external alterations to a church building or buildings, the churchyard, monuments, boundary walls and other features will require planning permission. This means that the local authority will become involved, and the important person here as far as archaeological implications to proposals are concerned is the archaeologist in charge of Development Control, often known as the Planning Archaeologist. This may or may not be the same person as the County Archaeologist and DAA. The Planning Archaeologist will give advice on applications for planning permission, and may recommend conditions regarding archaeology which will have to be fulfilled before planning permission is granted. They will be unlikely to give advice on mitigation (see Chapter 5), as their job is to respond to applications; the DAC should therefore be the first port of call for informal advice when a project is first mooted.

Similarly, if the church is within a conservation area, the local authority will have to be consulted in cases requiring change of appearance; the Conservation Officer will be the main point of contact in this case. Churchyards may also be affected by designation of all or part of them as Scheduled Ancient Monuments (SAMs), which will fall within the remit of the local English Heritage Inspector, though this is much rarer. The example of the church of St Kyneburgha, Castor, mentioned above is one such case. It should be noted that the church building itself cannot be so designated, which can lead to some confusion when works are planned which affect the floor levels and burial vaults within the church, for example.

Recent developments
From PPG to PPS

In recent years there have been various changes to the legal status of archaeology as expressed within the framework of secular planning processes, particularly Planning Policy Guidance notes (PPG) 15 (1994) and 16 (1990), although it is worth noting that these were being revised at the time of writing into a single Planning Policy Statement (PPS), part of sweeping changes to the planning system intended by the Government.

The need to take account of archaeology has formed part of the requirements of the faculty jurisdiction for many years, but it has been emphasized by the principles introduced by the PPGs. This means that parishes must be made aware of the need to take archaeology into account as part of their stewardship of buildings and churchyards. One of the key messages of PPG 16 and of recent CCC and DAC guidelines is to urge parishes to consider archaeology at an early stage in any planned works. This is equally true of works to a churchyard and to a church building.

PPG 16 also introduced the concept of 'developer pays' in relation to every kind of archaeological intervention. A consequence has been to cut most archaeologists loose from their bases within local authorities into the market place. This means there are now two types of archaeologists: curators, who monitor for free, and contractors, who do the work the curators specify for a charge. As far as a parish church is concerned, this means that the costs of any archaeological work required prior to or during programmes of maintenance or development in a church or churchyard, including archiving and, where appropriate, publication, are normally borne by the parish, though it may be possible in certain circumstances to apply for grant aid.

If the disturbance is to be caused as a result of necessary repairs which English Heritage is grant aiding, their grant may sometimes be extended to help offset archaeological costs. It may be worth making a representation to English Heritage anyway if the burden of archaeological work is obviously beyond the means of the parish, as indeed is recommended in Paragraph 25 of PPG 16.

There may be local sources of funding for archaeology, on which the DAA will be able to advise. Costs may be offset by using archaeologists as sub-contractors for the excavation element of the project, so ensuring that the final bill, though higher than if no archaeological intervention had been required, is not exorbitant.

The role of the architect is crucial. It is often difficult for parishes to see any tangible benefits from archaeological work, but easy for them to see the costs incurred by such work as an additional burden rather than as an integral, and necessary, part of the planning process. Early consultation between archaeologists and architects and the consequent mitigation of the archaeological implications of schemes of work will help to reduce friction, delay and financial distress for the parish, and here architects and archaeologists alike bear a great deal of responsibility. If there is a lack of communication, or indeed downright hostility or mistrust, between the archaeologists and the architect, only the parish is likely to suffer in the end.

Statements of Significance and Conservation Plans

A new requirement of parishes since the publication of the Faculty Jurisdiction Rules in 2000 is the preparation of a Statement of Significance when significant changes are proposed to a listed church. This should include an assessment of the archaeological significance of the church, churchyard and site and identify any threats to these inherent in the proposed works. Guidelines for producing these have been developed by several dioceses and also by the CCC; check with your DAC.

Conservation Plans are now expected from cathedrals, and guidelines for producing them are available from the Cathedrals Fabric Commission for England. Major churches would also be well advised to consider commissioning such a document, which should again contain a detailed assessment of the archaeological significance of the church, churchyard and site, obviously in more detail than is possible in a Statement of Significance. The preparation of these documents will be a useful exercise in itself, helping parishes to appreciate, understand and manage the buildings in their care, and the document itself will be a useful management tool.

fig. 11
The ruined church at Arborfield

The special case of ruined and redundant churches

A distinctive feature of most parts of the English countryside is the ruined church or chapel, often no more than a few lengths of wall or humps of rubble, but sometimes substantial remains spread over a large area, or surviving to some height (see fig. 11). Parishes often give no heed to these; however, in many cases these buildings have never been made formally redundant and are still legally under the care of the Church and the faculty jurisdiction, and therefore remain the charge of the church of the parish in which they lie. Should somebody be injured by falling masonry or toppling grave markers, the parish could be liable, as it is nominally responsible for the maintenance of the ruin and its curtilage. There will often be a churchyard attached and/or burials within the church itself. Should these be disturbed or desecrated, responsibility will again lie with the parish.

Recently, several parishes have attempted to resolve this anachronism by formally applying for redundancy for the ruined building under the Pastoral Measure (1983). This is far from an ideal situation in terms of the preservation of the ruins, as landowners are not often keen to shoulder the responsibilities of caring for what are in effect ancient monuments, primarily of landscape value and interest to the local community and of academic value to the historian and the archaeologist. The local PCC should take advice from the DAC, and it may well be advantageous to involve the secular parish council, the local authority, English Heritage and the landowner in an attempt to find a suitable solution.

Archaeologists have done much to quantify this problem. Professor Roberta Gilchrist undertook a survey of the ruined churches of England in 1989 for the CBA. This study established that there are at least 700 such ruined churches across the country. This does not include those of monastic institutions, which do not fall within the faculty jurisdiction unless parts of them were inherited by the parish for its worship needs, as is the case, for example, with the nave of Malmesbury Abbey. Such buildings will almost always be of high archaeological potential and significance, partly because they have not been altered since their abandonment and thus represent unadulterated examples of Romanesque and Gothic architecture, partly because they preserve evidence for liturgical and burial practices. Ruined churches are far from evenly distributed, with a notable concentration in Norfolk.

Redundant churches are a separate case, and will usually fall within the remit of the local authority once they have been removed from the faculty jurisdiction by the use of the Pastoral Measure. However, churchyards often remain consecrated and within the faculty jurisdiction, although no longer the responsibility of the parish. In cases where the Pastoral Measure has been used to achieve partial redundancy or reordering of the church buildings, it is wise to make use of the DAC's advice regarding the archaeological implications of any proposed works.

chapter 5
The mitigation of archaeological implications

The really quite complex system of planning and mitigation regarding archaeology can seem quite daunting to parishes. Add to this the archaeologist's love of specialist jargon, and it is quite easy to feel all at sea. This chapter is devoted to explaining what happens when changes to church buildings and churchyards are planned in regard to archaeology.

'We think our scheme may have archaeological implications: what happens next?' It cannot be emphasized too often or too strongly that the earlier consultation takes place, the less chance there is of unforeseen problems or costs arising. A 'mitigation strategy' devised with the help of the DAC will minimize difficulties, delay and cost, or may even help avoid them altogether.

Mitigation is the watchword here. It may be possible for the DAA to advise, for example, that if you re-route that drain run this way, or reduce the depth of the foundations, or reuse that blocked doorway rather than go through that medieval wall, there will be no or minimal archaeological implications; and so on. Consult your DAA through the DAC as early as possible, and you will find that they are ready and willing to help avoid archaeological costs to the parish as far as they are able within the constraints of a particular scheme.

Parishes will understandably often want to use the cheapest contractor, and it is difficult for DAAs and local authorities to insist on perhaps more expensive but more experienced and skilled contractors being employed; however, the results will almost always justify this often relatively minor increase in expenditure. Here architects have a responsibility to take advice from the DAA. Architects as the team leaders in most projects involving schemes of work with archaeological implications should avoid trying to cut corners with archaeology, especially in the important early stages of evaluation. It is no service to the parish to follow the mantra 'see no archaeology, hear no archaeology, pay for no archaeology'. The most unwelcome surprise for a parish, and often an expensive one, is to find late on in the progress of a scheme that extensive archaeological work is necessary, when too much money has already been spent to turn back and mitigation is no longer an option.

If archaeological work is unavoidable, it will be necessary to commission a brief for the work. A brief places the required archaeological work within the context of the overall project, outlines the work that will be required and sets out the research aims and objectives. It will normally be written by the DAA or a person approved by the DAA.

It is worth noting that, contrary to popular belief, archaeologists prefer to leave archaeological remains undisturbed and intact if there is no perceived threat to them, and, indeed, 'preservation *in situ*' of archaeological remains is the declared policy of the Government. From the Church's point of view, human remains should be left *in situ* wherever possible for practical reasons and out of respect for the dead. The following paragraphs outline the usual processes carried out to mitigate the archaeological impact of planned works to a church or its churchyard.

Archaeological appraisal

The DAA will be able to judge the likely impact of disturbances to any significant buried deposits or fabric through a process termed an 'appraisal'. The likely impact is dependent upon a number of factors. These include the area and/or depth of the anticipated intervention, its position in the church and churchyard and its relationship to any standing buildings and the known history of the area, including previous finds, documentary references and clues in the fabric of the church itself. Once the appraisal is complete, the DAA will make a recommendation as to the archaeological implications of the proposed work. There may be no need for any further archaeological involvement; if there is, there are several categories of archaeological recording which may need to be carried out before and/or during the proposed work.

Desk-based assessment

The next step after an appraisal is a thorough desk-based review of all existing archaeological information relating to the area under consideration. As the term implies, this is essentially a documentary exercise, with any work on site limited to non-intrusive investigation. No holes will be dug, no plaster stripped. The information gained at this stage may be adequate but, if not, it will be necessary to proceed to the next stage of investigation.

Watching briefs (precautionary monitoring)

For minor disturbances, for example a trench for a new pipe or for cables for floodlighting, or minor stone replacement or repointing, the recommended course of action is usually precautionary monitoring, often referred to as a 'watching brief'. This simply means having an archaeologist on site during the work to double-check that no significant historic deposits or features are being disturbed. This procedure is often appropriate for other types of disturbance, even as substantial as a small building extension, where this is being constructed on a 'raft' foundation to avoid deep digging. Watching briefs may sometimes, but by no means always, be provided at a modest, even nil, cost by local authority personnel or the DAA. In some circumstances it may be appropriate for an experienced amateur archaeologist to undertake this responsibility, but only with the prior approval of the DAC. A watching brief is unlikely to be an adequate response to larger works, which may require an archaeological evaluation or excavation and recording.

Archaeological evaluation

This usually takes the form of a small trial excavation to test the nature of the historic deposits, or a non-destructive survey using geophysics, for example by passing a small electric current underground to determine if walls, foundations or ditches are present, or a combination of these methods. The DAA will advise on suitable archaeological contractors who can carry out this work, providing if possible a choice of recommended individuals or organizations who, on invitation by the parish, may then tender for the work as set out in the brief.

Archaeological recording

Works which involve replacement of, or interference with, historic fabric may require recording before, during and after the work takes place. It must be emphasized that this can also be the case in like-for-like replacement of materials, for example a stone parapet, a monument or a roof timber, or part of a wooden screen. This can range from a high-quality photographic record to a comprehensive programme of detailed recording using the latest techniques, such as hand-drawn or photogrammetric stone-for-stone recording and electronic surveys. The recording of historic roof and bellframe structures often requires particular stringency, and may embrace such techniques as dendrochronology, Carbon-14 dating and three-dimensional computer-aided imaging techniques. There may be grants available for such scientific recording where the conservation or repair work is funded by bodies like the CCC, the Heritage Lottery Fund or English Heritage.

Archaeological excavation

Occasionally, because of the evident importance of the church building or site, archaeologists will need to undertake the work of a conventional excavation. Again, this should not be seen simply as an extra burden on parishes. The knowledge gained through such work can be of positive benefit to a parish, by adding to what is known about the history and development of the church and thus putting the building and its parishioners into a much wider context, and by aiding in the future management of the church building and churchyard.

Results: archiving, publication and dissemination

The results of archaeological work are not for the archaeologist to covet and hoard but must be distributed, if not necessarily published. Top of the list of recipients of this information must be in every case the customer, the parish. The archaeologists involved may be willing to give a short presentation on the results of their work to the local community, who will no doubt be interested in any discoveries that have been made. Too often the results of archaeological investigations, especially minor works, are simply filed away at the local SMR or museum and forgotten – so-called grey literature.

It is easy for parishes to see archaeological work as a waste of money and time, and the process simply as a hoop through which they must jump. To avoid this, archaeologists must ensure that they engage with and involve the parish, and vice versa. If this is done, everyone will benefit from 'church archaeology', which enriches, informs and empowers parishes and adds to the knowledge and cultural wealth of the nation.

Further reading

The legal framework

Ancient Monuments & Archaeological Areas Act (1979). Legislation relating to archaeological sites and monuments of national importance, protected as Scheduled Ancient Monuments. (This rarely affects churches themselves, but may be relevant to archaeological sites within churchyards, inscribed stones, churchyard crosses, or ruinous medieval buildings.)

William Dale, *The Law of the Parish Church*, 7th edn, Butterworths, 1998.

Norman Doe, *The Legal Framework of the Church of England*, Clarendon Press, 1996.

Lynne Leeder, *Ecclesiastical Law Handbook*, Sweet and Maxwell, 1997.

G. H. Newsom and G. L. Newsom, *Faculty Jurisdiction of the Church of England*, 2nd edn, Sweet and Maxwell, 1993.

Planning (Listed Buildings and Conservation Areas) Act (1990). Legislation relevant to churches and other associated buildings and structures classed as listed buildings or falling within conservation areas.

Planning Policy Guidance note 15 (PPG 15), Department of the Environment, 1994. Planning Policy Guidance: Planning and the historic environment.

Planning Policy Guidance note 16 (PPG 16), Department of the Environment, 1990. Planning Policy Guidance: Archaeology and planning.

General reading

J. Berrow (ed.), *Towards the Conservation and Restoration of Historic Organs*, CHP, 2000, esp. pp. 51–68.

J. Blair and C. Pyrah (eds), *Church Archaeology: Research Directions for the Future*, CBA Research Report no. 104, London, 1996.

The Churchyards Handbook, 4th edn, CHP, 2001, esp. pp. 82–93 and 122–6.

Thomas Cocke, Donald Finley, Richard Halsey and Elizabeth Williamson, *Recording a Church: An Illustrated Glossary*, CBA, 1996.

The Conservation and Repair of Bells and Bellframes, CHP, 2002.

Roberta Gilchrist, *Cencus of Ruined Churches*, vols 1 and 2, CBA, 1989.

Inside Churches: A Guide to Church Furnishings, National Association of Decorative and Fine Arts Societies, 1993.

Hilary Lees, *English Churchyard Memorials*, Tempus Publishing, 2000.

R. Morris, *Churches in the Landscape*, Dent, 1989.

Harold Mytum, *Recording and Analysing Graveyards*, CBA, 2000.

New Work in Historic Places of Worship, English Heritage, 2003.

W. Rodwell, *The Archaeology of the English Church*, Batsford, 1981.

W. Rodwell, *The English Heritage Book of Church Archaeology*, Batsford, 1989.

Organizations able to give advice

Amenity Societies
The Amenity Societies (of which the CBA is one) will be interested in offering advice and commenting on proposals affecting church buildings of particular periods.

Ancient Monuments Society
St Ann's Vestry Hall
2 Church Entry
London EC4V 5HB
www.ancientmonumentssociety.org.uk

Council for British Archaeology
Bowes Morrell House
111 Walmgate
York YO1 2UA
Tel: 01904 671 417
www.britarch.ac.uk/cba

Council for the Care of Churches
Church House
Great Smith Street
London SW1P 3NZ
Tel: 020 7898 1866
www.cofe.anglican.org

Diocesan Advisory Committees (DACs)
DACs should be contacted prior to any work being undertaken and can offer advice on archaeological matters through the DAA. A list of DAC addresses can be found at www.churchcare.co.uk

English Heritage
23 Savile Row
London W1X 1AB
Tel: 020 7973 3000
www.english-heritage.org.uk

Or contact English Heritage's regional offices (details on their web site)

The Georgian Group
6 Fitzroy Square
London W1T 5DX
Tel: 020 7529 8929
www.georgiangroup.org.uk

Society for the Protection of Ancient Buildings
37 Spital Square
London E1 6DY
Tel: 020 7247 5296
www.spab.org.uk

The Twentieth Century Society
70 Cowcross Street
London EC1M 6EJ
Tel: 020 7250 3857
www.c20society.demon.co.uk

The Victorian Society
1 Priory Gardens
Bedford Park
London W4 1TT
Tel: 020 8994 1019
www.victorian-society.org.uk

Index

abbey churches
  Malmesbury Abbey, screen 15
  Romsey Abbey, ringing
    chamber 11
  see also churches
advisory organizations 43
aerial photography 4
altars 14, 16
Amenity Societies 30, 43
Ancient Monuments &
  Archaeological Areas Act
    (1979) 41
Ancient Monuments Act (1882) 30
appraisals 37
archaeological briefs 37
archaeological evaluations 38
archaeological implications of
  renovation works 6–29
archaeological recording 39
archaeology
  definition of church archaeology 3
  importance 1–2
  legal status, in relation to planning
    systems 32
  scope 3–4
Archbishops' Council, Cathedral
  and Church Buildings Division
    29, 30
archiving 39–40
artefacts, in churchyards 26–7

belfries 11
bellframes 10–11, 39
bells 10
benches, dating 14
buildings control legislation 30

cathedral churches
  Conservation Plans 33
  Norwich Cathedral, Dispenser
    reredos 17
  Salisbury Cathedral, roof 12
  Winchester Cathedral, organ 16
  see also churches
Cathedrals Fabric Commission
  for England 33
CBA (Council for British
  Archaeology) 3
CCC *see* Council for the Care
  of Churches
Channel Tunnel Rail Link 28
chapels
  chantry chapels 7–8, 9
  side chapels 16
chests, medieval 14
*Church Archaeology: Its Care and
  Management* (CCC) 27–8
church archaeology *see* archaeology
church buildings
  fabric 4, 7–9
  interiors, reordering 21
  legal requirements for proposed
    developments 30–3
  redundant churches, legal
    responsibility for 35
  ruined churches, legal
    responsibility for 34–5
churches
  Asheldham (Essex), priest's
    house 24
  Aylsham, St Michael, ringing
    chamber 11
  Bosham, Holy Trinity, spire 12

Index

Bradwell-on-Sea (Essex), St Peter 7
Brancepeth, St Brandon, grave slabs 19
Brent Eleigh, St Mary, reredos 16, 17
Cambridge, King's College, clock 10
Castor, St Kyneburgha
 churchyard designated as SAC 31
 Roman site 24
Caversfield, St Lawrence, Gargate bell 10
Chesterfield, St Mary, spire 12
Chichester, St Mary's Hospital, screen 14
Claughton, St Chad, bell 10
Crondall, All Saints, housing in churchyard 25
Denton, St Lawrence, timber frame 12 fig.
Doddinghurst, All Saints, belfry 11
Dunsfold, All Saints, pews 14
Fulbourn (Cambridgeshire), St Vigor with All Saints, churchyard layout 25–6
Gamlingay, St Mary the Virgin, pews 14
Greensted-iuxta-Ongar, St Andrew, date 12
Hardham, St Botolph, bell tower 10
Houghton-on-the-Hill, St Mary, wall paintings 17
Hull, Lowgate, St Mary, bellframe 11
Ifield, St Margaret, funerary monuments 21
Kempley, St Mary
 furnishings and fittings 14
 roof and wall paintings 11–12
Lambley (Diocese of Southwell), Holy Trinity 8–9
Lincoln, St Michael, churchyard 24
Lincoln, St Peter, churchyard 24
Livermere, St Peter
 funerary monuments 20 fig.
 tower 8 fig.
London, St Pancras Old Church, churchyard 28
Malmesbury Abbey, parochial use of the nave 35
Nottingham, St George, organ 16
Old Radnor, organ 16
Onibury, St Michael, bellframe 11
Oxford, St Mary the Virgin, turret clocks 10
Rivenhall, St Mary and All Saints, Roman site 24
St Bartholomew the Less, bellframe 10–11
Sherburn, St Hilda, Saxon carvings 19 fig.
Skelton-in-Cleveland, All Saints, furniture and fittings 14
Stanton Harcourt, St Michael, rood screen 15
Temple Hill (Dartford), St Edmund the Martyr, churchyard 25
Thornam Parva, St Mary, retable 16–17
Troston, St Mary, wall paintings 17 fig.
Warburton (Cheshire), St Werburgh, timber-frame 12
Warmington, St Mary the Virgin, pews 14
Weaverthorpe, St Andrew, funerary monuments 21 fig.
Whitby, St Mary, furniture and fittings 14
Wintringham, St Peter
 scratch dial 22 fig.
 screen 15
 stained glass 18
see also abbey churches; cathedral churches; churchyards

Index

churchyards
  archaeology 24–6
  artefacts 26–7
  legal requirements for
    developments 31, 32
  recording of monuments 19
  reordering and the treatment of
    human remains 27–9
  reordering and churchyard
    plans 23
  ruined churchyards, legal
    responsibility for 34–5
  see also churches
Clement, William (clockmaker) 10
clocks, turret clocks 10
conservation areas 31
Conservation Officers 31
Conservation Plans 33
consultations, legal requirements
  30–1, 36
contractors, cost levels 36
costs 36–7, 39
  defrayment of archaeological
    costs 32–3
Council for British Archaeology
  (CBA) 3
Council for the Care of Churches
  (CCC) 10, 27–8, 30, 33
  grants 39
county archaeologists 26
cross-slabs 18–19

DAA *see* Diocesan Archaeology
  Advisors
DAC *see* Diocesan Advisory
  Committees
dendrochronology (tree-ring
  dating) 4
  dating of furnishings and
    fittings 14
  impact on dating of bellframes
    10–11
  impact on dating of roofs 11–12
  impact on dating of structural
    timbers 12, 13 fig.
desk-based assessments 38
'developer pays' concept 32
developmental works, legal
  requirements 30–1
Diocesan Advisory Committees
  (DAC) 10, 11, 18, 22, 30, 33, 43
  Advise to Diocesan Chancellors 31
  consultations with 35
  and mitigation strategies 36
Diocesan Archaeology Advisors
  (DAA) 22, 26, 30
  advice on archaeological costs 33
  and mitigation strategies 36,
    37, 38
Diocesan Chancellors
  discretion over treatment of
    disinterred human remains 29
  granting of faculties for
    developmental works 31
Diocesan Record Offices, as
  depositaries for churchyard
  plans 23
DNA study 4
  use during disinterment of human
    remains 28
doorways 8, 9, 14, 15
drainage, improvement, effects on
  churchyards 24
Drewett, Cecil 11

East Anglia, church architecture 9
Ecclesiastical Exemption 30, 31
English Heritage 29
  development of Grade I and Grade
    II* churches 30
  grants 32, 39
excavations 39
  Lambley, Holy Trinity 9

Faculty Jurisdiction Rules (2000) 33
faculty jurisdictions 27–8, 30, 31

Index

fairs, medieval 24
Federation of Family History Societies 23
fittings, and furnishings 14–18
floors 12–13, 21, 31
fonts 16, 18
funerary monuments 18–22
  memorial brasses 13
  recording 23
  see also graves; human remains
furnishings, and fittings 14–18

galleries 15–16
geophysical surveys 4, 23
Gilchrist, Roberta 35
glass, stained glass 18
grants 32, 39
grave markers 23, 24, 27, 34
grave vaults 5, 13, 22, 31
graves 4, 13, 25, 26, 34
  see also funerary monuments

heating systems, renewal 13
Heritage Lottery Fund, grants 39
Historic Environment Records (HERs) 25
housing, remains in churchyards 24, 25
human remains
  excavation 3
  preservation *in situ* 37
  recording of funerary monuments 20
  reordering of churchyards 25, 27–9

incumbents, ownership of artefacts discovered in churchyards 26
Iron Age sites 25

Knibb, Joseph (clockmaker) 10

legal requirements and proposed developments 30–1

legal responsibilities, for ruined churches 34–5
local authorities 31, 38
local history societies 23

markets 24
Meet the Ancestors 28
mitigation 31, 36–40
monuments *see* funerary monuments

New Work in Historic Places of Worship (English Heritage) 30
Norfolk, ruined churches 35

organs 16
Otho (Legate in England of Pope Gregory IX) 30

Parochial Church Councils (PCC) 18, 30
  maintenance of ruined churches and churchyards 35
  ownership of artefacts discovered in churchyards 26
Pastoral Measure (1983) 35
pews 13, 14
piscinae 16
Planning Archaeologists 31
Planning (Listed Buildings and Conservation Areas) Act (1990) 41
planning permission 31
Planning Policy Guidance (PPG) notes 32
  PPG 15 (1994) 32
  PPG 16 (1990) 32
Planning Policy Statements (PPS) 32
planning systems, legal status of archaeology 32
precautionary monitoring (watching briefs) 38
preservation *in situ* 37
pulpits 14

47

Index

renovation works, archaeological implications 6–29
reredos 16, 17
retables 16–17
Roman sites, underlying churchyards 24, 25
rood screens 15, 16
roofs 11–12
  archaeological recording 39

Scheduled Ancient Monuments (SACs) 31
screens 14–15
Shakespeare, William, grave marker 27
Sites and Monuments Records (SMRs) 25, 26–7
Society for Church Archaeology 4
Society of Genealogists 23
spires 12
squints 16
Statements of Significance 33

sundials 22

Time Team 28
Tompion, Thomas 10
towers 8 fig., 9
  and bellframes 10–11
tree-ring dating *see* dendrochronology

vaults *see* grave vaults

wall paintings 17–18, 22
wall panels 14
walls, tracing 5
watching briefs (precautionary monitoring) 38
William I 'the Conqueror' 30
windows 8, 22
  stained glass 18
wood
  structural timbers 12, 13 fig.
  see also dendrochronology

48